Side by Side

SIDE BY SIDE

All sorts of animals live close to other kinds of animals. Some are partners and help each other, while others are parasites and harm the animals they live with. Plants can be partners with animals, but animals that live inside plants do not help the plants they live in and often destroy them. Plants can also be attacked by other plants.

Side by Side tells us about some of these weird and wonderful relationships, from shrimps that clean between the teeth of fish to insects that make a meal out of human blood.

For additional details about each picture, you can turn to More Facts at the end of this book.

Acknowledgements

Grateful acknowledgement is made to:
Tony Allen: page 15
Animals Animals: pages 8, 14 (Richard Kolar)
Kathie Atkinson: page 13 (bottom)
George Bernard: page 12 (bottom), page 29 (top)
Raymond Blythe: page 28
J. A. L. Cooke: pages 17, 26
D. J. DeVries: page 21 (bottom)
Fredrik Ehrenstrom: page 9
Michael Fogden: pages 10, 11
Terry Heathcote: title page
Zig Leszczvnski: page 7
Mantis Wildlife Films: pages 24, 25 (Jim Frazier)
Raymond Mendez: page 12 (top), back cover
Ben Osborne: page 22
Stan Osolinski: front cover, page 18
Peter Parks: page 5, 19 (top)
D. J. Saunders: page 20, 23
David Shale: pages 4, 19 (bottom)
Tim Shepherd: pages 16, 29 (bottom)
D. H. Thompson: page 21 (top)
P & W Ward: pages 13 (top), 27
Neville Zell: page 6

British Library Cataloguing in Publication Data

Side by Side
  1. Organisms. Symbiosis
  I. MacQuitty, Miranda
  574.5'2482

  ISBN 0 233 98291 4

First published in 1988 by
André Deutsch Limited
105–106 Great Russell Street, London WC1B 3LJ

ISBN 0 233 98291 4

Fallow Bucks

Side by Side

Oxford Scientific Films
edited by
MIRANDA MACQUITTY

ANDRE DEUTSCH

USE MY WEAPONS

Sea anemones and jellyfish catch and kill small fish with their stinging tentacles. But some fish are not hurt by these weapons. They even hide among the tentacles from larger fish, that might attack them.

Lion's Mane Jellyfish. A shoal of baby fish keeps close to this jellyfish. Its stinging tentacles will keep predators such as larger fish away.

Pink Jellyfish.

A young fish has a safe home under the umbrella of this jellyfish.

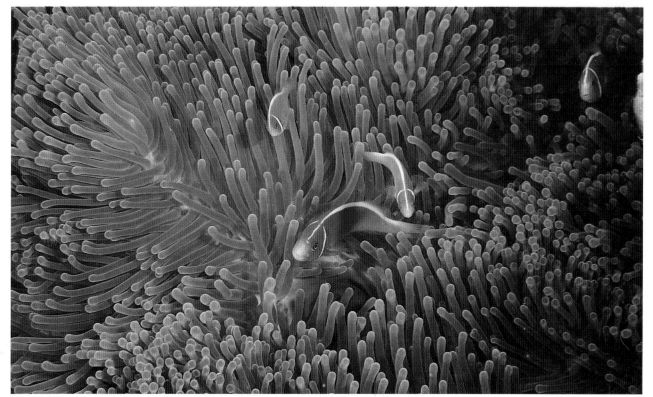

Sea Anemone. These coral reef anemone fish nestle among the tentacles of a giant anemone.

HITCHING A RIDE

Some animals are hitchhikers and hang onto other animals for a free ride.

Remoras. By sticking onto a manta ray these lazy fish don't need to swim themselves.

Sea Anemone. Riding along on the shell of a hermit crab gives this anemone a chance to eat the leftovers from the crab's dinner.

BARNACLES ON BOARD

Barnacles are usually found stuck onto rocks on the seashore. But some barnacles have mobile homes on other animals.

Whale Barnacles. This is not a rock but the head of a whale. The light brown patches are bunches of barnacles.

Acorn Barnacle. This barnacle, probably so-called because it looks rather like an acorn, is riding along on a seashore crab.

JUNGLE TRIOS

Sloths laze about in tropical forests hardly moving at all. Both animals and plants live among their fur.

Plants on Sloths. This sloth mother looks green because she has tiny plants growing in her fur.

Moths on Sloths. This sloth is sunbathing. Can you see that there are small moths walking along its back?

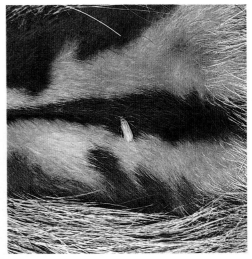

Sloth Moth.
In this close-up picture you can see one of these moths quite easily on the sloth's fur.

ANIMALS ON ANIMALS

Some animals have bloodsuckers living on them. They are called parasites, and usually feed on their hosts without killing them.

Tick. This tiny tick is crawling over a huge python snake. It will bite the snake between its scales.

Louse Fly. This fly lives on birds. It will bite the bird's skin beneath its feathers.

Mites. The red blobs on this harvestman are baby mites. The mites get a free ride and feed on the harvestman at the same time.

Fish Louse. This large louse is living on a damsel fish.

ANIMAL INVADERS

Some animals live inside other animals. These invaders feed on their victims.

Braconid Wasps.
This caterpillar's skin is covered with white wasp cocoons. The wasp larvae ate the insides of the caterpillar before turning into cocoons.

Fly and Field Cricket. The fly is squirting out its tiny maggots onto a cricket. The maggots will burrow into the cricket to feed.

Fly Pupa in a Field Cricket. By now the fly maggots have eaten up the cricket's insides. One maggot has crawled out of the cricket and turned into a brown pupa.

POSTING DISEASES ON PEOPLE

Sometimes tiny animals feed on us. When they pierce our skin they can give us diseases.

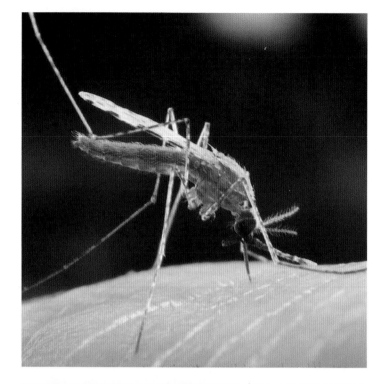

Mosquito.
This mosquito has just alighted on a finger, pierced the skin and begun to feed.

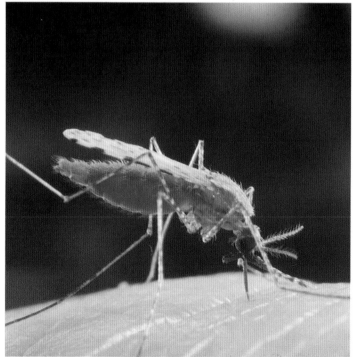

Mosquito Sucking Blood.
The mosquito's body is now full of red blood.

A Soft Tick Before a Meal.
This tick is sitting on a human arm.

A Soft Tick After a Meal.
The tick has soft and stretchy skin,
so as it sucks blood it swells up.

CLEANER SERVICE

Some fish and birds pick harmful parasites off other animals and clean up their wounds. The cleaners serve their customers well and get a good meal by eating the parasites.

Red-billed Ox-peckers. The birds perching on this African buffalo pick off ticks and other parasites living on its skin, and even in its nose.

Cleaner Fish. Painted sweet lips fish (fishes of coral lagoons, so-called because of their soft-looking lips) are waiting their turn to be cleaned by the small striped cleaner fish already at work on the nearest fish.

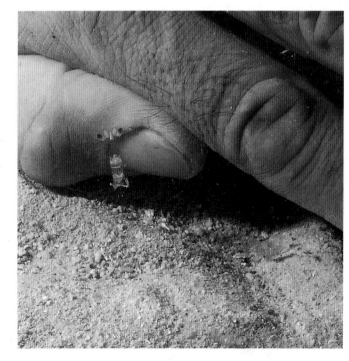

Cleaner Shrimp.
This shrimp usually cleans fish but here it is tweaking at a scuba diver's hand.

ANTS AS PROTECTORS

Ants make very good guards since they are well armed. They can bite attackers with their powerful jaws and burn them by spraying acid.

Bathing with Ants. The jay has ants crawling between its feathers. It uses the ants' acid spray to get rid of lice and other parasites that are biting it.

Caterpillars. Ants look after caterpillars because they make sugary droplets that the ants like to eat.

These ants are guarding the caterpillars of an Australian butterfly.

These two ants are drinking the sugary droplets made by this Costa Rican caterpillar.

BIRD BULLIES

Birds do not always live happily side by side. Some birds steal food and others take over nests.

Skua.
The skua scared this albatross chick into throwing up its dinner. Now the skua is eating the free meal.

Cuckoo. Two reed warbler parents feed a cuckoo chick twice their size. The crafty cuckoo pushed all the reed warblers' own eggs out of the nest over a week ago.

ANIMALS IN PLANTS

Some animals live inside plants, where they have shelter and plenty to eat.

Leaf-mining Fly. A fly maggot made this pattern by tunnelling inside the leaf.

Leaf-mining Fly. Close up, you can see the maggot eating the leaf.

PLANT WARTS

Plants have warts, much as we do, but they are called galls, and animals live inside them.

Spangle Galls. Larvae of tiny wasps are living inside the yellow bumps or galls on the underside of this oak leaf.

Robin's Pincushion.
These furry galls, or robin's pincushion, grow on wild rose bushes.

Wasp Grubs and Pupae. A robin's pincushion, cut open to show the animals living inside it.

PLANT ENEMIES

Mean Mistletoe

Mistletoe grows on many kinds of trees and steals water and food from them.

In the winter, when the trees have lost their leaves, you can see clumps of mistletoe among their branches. We use mistletoe to decorate our homes at Christmas.

A branch of mistletoe on a fruit tree.

A tree branch has been sliced open to show how the mistletoe grows through the brown bark into the wood.

MORE FACTS

USE MY WEAPONS (pages 4 & 5)

Sea Anemone. Although the anemone fish are safe among the anemone's stinging tentacles, other fish touching the tentacles are stung. Young anemone fish at first bring only a small part of their bodies in contact with the tentacles. After repeatedly touching the tentacles the young fish realize they are not harmed and can settle down to live in the anemone. Giant anemones can be over one metre wide so there is plenty of room for several fish to live in one anemone.

Pink Jellyfish. This jellyfish is from the Great Barrier Reef of Australia. Many kinds of jellyfish are found on the Great Barrier Reef and some, like this one, protect young fish. These fish are not harmed by the jellyfish's stinging tentacles, but larger fish trying to eat the young fish will be badly stung.

Lion's Mane Jellyfish. This jellyfish is also from the Great Barrier Reef of Australia. It can grow up to two-thirds of a metre across and often has a whole shoal of young fish swimming among its tentacles. The young fish are protected from larger predatory fish by the jellyfish's stinging tentacles but are not hurt themselves. Once the fish mature they leave the jellyfish and fend for themselves.

HITCHING A RIDE (pages 6 & 7)

Remoras. Rather than swimming themselves, these fish ride along on other animals such as rays, sharks and turtles. They cling on with a sucker which is on top of their heads. Remoras riding on sharks will eat scraps of food missed by these fearsome fish as they tear up their prey.

Sea Anemone. Sea anemones often clamber aboard the shells of hermit crabs so they can grab morsels of food from the crab with their sticky tentacles. Sometimes hermit crabs pick up sea anemones and place them on their shells because the anemone's stinging tentacles help ward off attackers, such as fish.

BARNACLES ON BOARD (pages 8 & 9)

Whale Barnacles. These barnacles are living on the California Gray whale. The barnacles feed on plankton which they filter out of the seawater as the whales swim along. The barnacles travel hundreds of miles each year as the California Gray whales migrate along the entire length of the Pacific coast of North America, from their feeding grounds near the Arctic Circle to their breeding grounds in the warm lagoons of Baja, California. The barnacles release their young into the ocean where they must find another whale to settle on if they are to survive.

Acorn Barnacle. This kind of barnacle usually lives on hard surfaces such as rocks in the sea. Here one barnacle has stuck onto the shell of a seashore crab. You can see the delicate feeding legs which the barnacle uses to filter particles and plankton out of the water to eat. When the crab moults the barnacle will be cast off with the crab's old shell and may be killed if the shell is broken up by the pounding waves.

JUNGLE TRIOS (pages 10 & 11)

Plants on Sloths. Tiny plants called blue-green algae live in the shaggy fur of sloths. The algae grow in grooves along the length of each of the sloth's long hairs. In wet weather the algae grow well and the sloth's fur looks bright green. This helps the sloth to be camouflaged in its leafy surroundings.

Moths on Sloths. Many moths can live in the fur of the three-toed sloth. The female moth lays her eggs in the sloth droppings which the sloth carefully buries in the ground. The grubs feed on the droppings, then pupate and turn into moths which fly off to find another sloth to live on.

ANIMALS ON ANIMALS (pages 12 & 13)

Some animals (called parasites) live on other animals (called hosts) and harm them. Parasites either suck their host's blood or eat its skin and other tissues. Parasites also live inside animals (see Animal Invaders). Although hosts are hurt by parasites, they are rarely killed because parasites need to eat them for food.

Tick. The tick feeds on the python's blood using its sharp mouthparts to pierce the snake's skin. Ticks usually feed on the snake's back or sides where they are unlikely to be knocked off as the snake crawls along. Ticks are found on other reptiles such as lizards, as well as on birds and mammals.

Mites. The young mites (larvae) catch onto any harvestman wandering by. As the mites hitch a ride they suck up the harvestman's body fluids. When the mites have had their fill they drop off and then moult. Adults of this kind of mite are not parasites but predators of insects. Harvestmen are relatives of spiders.

Louse Fly. This blood-sucking parasite feeds on woodland birds such as pigeons, owls and cuckoos. The fly's body is flattened allowing it to move easily between the feathers and it has strong claws to hang onto them. Unlike a true louse the fly has wings so it can leave its host bird.

Fish Louse. The fish louse lives on the head of damsel fish from the Australian Great Barrier Reef. It feeds on the fish's blood and tissues. Damsel fish infected by lice grow slowly and small damsel fish may even be killed by lice. The fish louse's cousins are woodlice which live on land and feed on rotting wood and leaves.

ANIMAL INVADERS (pages 14 & 15)

Animals that live inside other animals are also called parasites. Some spend all their lives within their hosts while others, like the examples here, are only parasites when they are young. Unlike 'Animals on Animals' the parasites here kill their hosts.

Braconid Wasps. Over fifty wasp grubs fed on this sphinx moth caterpillar. The grubs eat the caterpillar's fat reserves first, leaving the essential body organs for last so the caterpillar is kept alive for as long as possible. When the grubs finish feeding they break out of the caterpillar and pupate in white cocoons. Adult wasps eventually emerge from the cocoons. Many butterfly and moth caterpillars are attacked and killed by wasps in this way.

Fly and Field Cricket. The female fly does not lay eggs but instead produces active maggot young. She lands on the cricket for only a few seconds to squirt out her tiny maggots onto its back. Then the maggots burrow into the cricket to feed. When the maggots are ready to pupate they break out of the cricket. Other kinds of parasitic flies are beneficial to man as they kill caterpillar pests on crops.

POSTING DISEASES ON PEOPLE (pages 16 & 17)

Mosquito. To draw blood the mosquito inserts its needle-sharp mouthparts into a blood capillary just below the surface of the skin. (It doesn't actually bite.) First the mosquito injects saliva to stop the blood from clotting and blocking its thin feeding tube. This substance often irritates our skin causing the itchy mosquito 'bite'. Only the female mosquito feeds on blood which she needs to produce ripe eggs. The male mosquito feeds on plant juices. Mosquitoes can carry diseases such as malaria and elephantiasis which they transmit when they feed on people.

Soft Tick. This parasite is found in Africa and carries the deadly disease, African relapsing fever. During the daytime ticks hide in crevices in the walls of huts and houses. At night they come out to feed on people while they sleep. They will also bite domestic animals like pigs, goats, sheep and dogs.

CLEANER SERVICE (pages 18 & 19)

Cleaner Fish. Cleaner fish live on coral reefs. The broad black stripe along its body is an advertisement, telling other fish it is a cleaner and therefore not to be eaten. The cleaner fish stays at one spot on the reef called a 'cleaner station' so its customers know where to find it. A fish customer presents the part of its body that needs attention and hardly moves while the cleaner nibbles off parasites and diseased skin. The cleaner even goes inside its customers' mouths and between gills to clean.

Cleaner Shrimp. When one of Oxford Scientific Films' cameramen offered his hand to a cleaner shrimp he was most surprised to be treated to a brush-up just like a big fish. The shrimp uses its tiny pincers to do the cleaning. Normally it picks off parasites and infected skin from fish. Like the cleaner fish, the shrimp can safely go inside the mouths of fishes without fear of being eaten since the fish are grateful for having parasites removed. When the shrimp's work is done it shelters with a sea anemone.

Red-billed Ox-peckers. These birds clean a wide range of animal customers such as buffaloes, rhinoceroses, giraffes and large antelopes. The ox-peckers help their customers not only by ridding them of parasites but also by warning them of approaching danger. While perching on their customers, the birds keep watch for predators such as lions.

ANTS AS PROTECTORS (pages 20 & 21)

Ants and Caterpillars. Ants protect many kinds of butterfly caterpillars in return for the sugary droplets the caterpillars produce from their honey-glands. The ants encourage the caterpillars to secrete the sugary substance by stroking them with their antennae.

Some caterpillars, like this one from Costa Rica, actually have tube-like honey-glands which they put out from the sides of their body when the ants want to drink. If you look closely you can see that each ant is drinking from the tip of a honey-gland, rather as we drink through a straw.

Caterpillars of the Australian Genoveva Azure butterfly live in ants' nests at the bottom of eucalyptus trees infested with mistletoe. At night, when the risk of being eaten is less, the caterpillars leave the nest to feed on the mistletoe.

Bathing with Ants. Many kinds of birds use the formic acid produced by ants to get rid of parasites. The jay uses living ants and spreads out its wings on the ground to encourage the ants to climb aboard. Other birds pick up ants and crush them onto their skin where they are being bitten by parasites. After preening these birds sometimes eat the ants. Birds will also preen with other kinds of insects and even with pungent pickles and onions picked out from garbage.

BIRD BULLIES (pages 22 & 23)

Skua. Skuas often get free meals by bullying other birds. In South Georgia in the Antarctic the albatross chick was alarmed by the skua's mock attacks and threw up its food. Skuas will harass birds such as gulls while they are in flight, forcing them to give up their food.

Cuckoo. When a female cuckoo is ready to lay her eggs she finds another bird's nest. First she normally steals one of the eggs present in the nest, then she lays her own egg. When the nest owners return they continue to incubate the eggs, including the cuckoo's, which is only slightly larger than the others. The cuckoo usually hatches out first and pushes the other eggs out of the nest. Finally only the cuckoo chick remains in the nest and the owners work hard to find it enough food.

ANIMALS IN PLANTS (pages 24 & 25)

Leaf-mining Fly. These flies can be identified by the shape of their tunnels. The fly maggot never goes back through an old tunnel so it always feeds on a fresh part of the leaf. The young of other insects such as moths and beetles are also leaf-miners. Leaves can be extensively damaged by miners and may die.

PLANT WARTS (pages 26 & 27)

Plants make galls around the insect eggs laid inside them. When the insects hatch they feed on the gall instead of damaging other parts of the plant.

Spangle Galls. These are caused by a wasp. In the summer the female wasp lays her eggs beneath the surface of the oak leaf. The leaf forms a gall around each of the eggs and when the eggs hatch the grubs eat the inside of the galls. In the autumn the galls drop to the ground where the grubs continue to feed safely inside them. Winged wasps burrow out of the galls in the spring.

Robin's Pincushion. This gall is often found on wild roses and is caused by a wasp. In spring the female wasp lays her eggs on unopened leaf buds. Instead of growing into normal leaves the bud forms a pincushion gall around the wasp's eggs. When the eggs hatch the grubs feed on the inside of the gall. The wasps pupate within the pincushion and crawl out as winged adults the following spring.

PLANT ENEMIES (pages 28 & 29)

Mean Mistletoe. This plant grows on trees instead of in the soil. It sucks up water and nutrients from the tree's sap but since it has green leaves it also makes some of its own food using sunlight by a process called photosynthesis. Mistletoe has white berries which are eaten by birds. The seeds inside the berries are not digested by the birds and often land on another tree in the bird's droppings. Here the seed germinates and a new mistletoe plant grows. Sometimes birds get the sticky juice of the mistletoe berries around their beaks. As they clean their beaks on the bark of trees they plant seeds in crevices in the bark. You can start your own mistletoe plant by making a small cut in the bark of an apple tree and squashing a white mistletoe berry into the cut.

PRINTED IN BELGIUM BY

proost
INTERNATIONAL BOOK PRODUCTION